"David Rabb engages young readers in all aspects of military life, and shows the vital role our veterans play in protecting this country. This wonderful book is a must for our kids, and thought provoking for their parents as well."

— Richard North Patterson, New York Times bestselling author of *In The Name of Honor*, *The Spire*, and *Balance of Power*

"As a child psychologist, I am so pleased to see Col. Rabb create a book that helps children understand the military culture."

— Barbara Van Dahlen, Ph.D., Founder and President of *Give an Hour*

"... a beautifully illustrated, compelling book specifically developed for our nation's young people. Our children need to be familiar with the sacrifices our veterans make, and this important book helps to bridge that gap."

— Keith Armstrong, Clinical Professor of Psychiatry; Co-author of *Courage After Fire: Coping Strategies for Troops Returning from Iraq and Afghanistan and Their Families*

"I highly recommend *FROM A TO Z – WHAT A VETERAN MEANS TO ME* to any child, family or adult interested in learning about the dedicated service of those who risk their lives for our safety, protection and peace."

— Joe Bobrow, Zen teacher, Founder and President of the *Coming Home Project*

"Illuminating. This unusual little book brought tears to my eyes ... It also brought joy to my heart through its luminous illustrations and its simple but clear message about who veterans are and what matters to them."

— Parvati Dev, PhD, FACMI, President of *Innovation in Learning Inc.*

"This book really takes an interesting perspective. It not only honors our veterans, but it explains in very understandable terms why it is so important to do that! It will really be a wonderful addition to schools' media centers and a treasured resource for teachers."

— Catherine Ost, Gifted and Talented Education Specialist

"This book is a wonderful example of how *everyone* can make a difference."

— Lisa Rosenthal, Founder of *Vet Art Project*

"We must never forget the debt we owe our veterans, and we must teach our next generation the culture of true patriotism. This book does that."

— John J Waickwicz, Rear Admiral United States Navy (Ret.), Former Commander Naval Mine and Anti-Submarine Warfare Command

"…a wonderful tool for children and their parents to learn about this complex institution in a simple, straightforward, and eye-catching read."

— Amy N. Fairweather, Director of Policy, *Swords to Plowshares*; Program Director *Coalition for Iraq and Afghanistan Veterans*

"This charming little tome teaches American children the cultural mores of our military veterans. Clever, moving, and chock full of little details that kids should know."

— Joyce Faulkner, President of *Military Writers Society of America*

"*FROM A TO Z* is a patriotic, reassuring, and understandable look at veterans for civilians and military alike."

— SSG Jay Wilkerson, Wounded Warrior

"Reading *FROM A TO Z – WHAT A VETERAN MEANS TO ME* is one small way to honor our veterans by providing America's children the opportunity to be touched by what exists and what has always existed in the heart and soul of our nation's heroes."

— Kathleen Cutshall, Licensed Clinical Social Worker

"This is a lot more than an alphabet book for young children; it's a window into how one contemporary American career soldier wants our youth to understand and appreciate the military life and its responsibilities. A great gift for boys and girls — anytime."

— Dana Hendrickson, President and Founder of *Rebuild Hope*

From A to Z
What a Veteran Means to Me!

A Look into the United States Armed Forces' Culture and Veterans Community

Written by David Rabb

Illustrated by Isha Gupta & Edited by Gaya Gupta

NAVIGATOR BOOKS
SAN DIEGO, CALIFORNIA

From A to Z
What a Veteran Means to Me!

Text Copyright © 2010 by David Rabb

Illustrations Copyright © 2010 by Isha Gupta

All rights reserved. No part of this book may be used or reproduced by any means, graphic, electronic, or mechanical, including photocopying, recording, taping or by any information storage retrieval system without the written permission of the author or publisher, except in the case of brief quotations embodied in critical articles and reviews.

Navigator Books
www.navigator-books.com

The illustrations in the book were created by an 11 year old girl who has no formal connection with the United States military. Although some of the pictures may lack in technical accuracy, they are beautiful and heartfelt portrayals of the U.S. Armed Forces, as seen through the eyes of a child.

The Veteran's Day illustration (V is for Veteran's Day) was adapted from an image of the 2008 Veterans' Day Poster. The original image is the property of the Department of Veterans' Affairs, and has been released to the public domain.

This book has been reviewed by representatives of the Department of the Army Staff Judge Advocate. The views expressed herein are those of the author, and do not necessarily represent the views of the Department of Defense, or its Components.

ISBN: 978-0-9890026-2-2

Community heals. Isolation kills.

http://fromatozveteransbook.com

Printed in the United States of America

This book is dedicated to the unyielding spirit and steadfast will of the men and women of the U.S. Armed Forces who came before us, who are currently serving or who will join the ranks in days and years to come. On your shoulders America stands. May God bless you and your family.

Freedom & Joy,

— *David Rabb*

I would like to dedicate this book to my family, whose support and encouragement made it possible for me to be a part of this.

I would also like to dedicate this book to our Veterans and their families.

— *Isha Gupta*

Foreword

Children know more about our world today than we ever did at their age. They are exposed to a constant barrage of around-the-world and around-the-clock news reports of battles, wars, and terrorist's attacks. Yet, there still remains that innocence of youth which doesn't fully understand what our military does, or how that impacts his or her own lives. This book opens that discussion through an informative approach using simple text and emotionally moving illustrations.

This book teaches children respect, patriotism, and honor—virtues that are rarely cultivated in our culture and society; and it is done in a very simple way—much like learning the A, B, C's.

I have personally been working, for well over three decades, with veterans who are dealing with Post-Traumatic Stress Disorder. So, it is worth noting that this wonderfully designed and written children's book is actually going to help fund a non-profit organization that deals with PTSD. Not only that—but it will educate young readers in a way that they would not get through any normal grade school classroom curriculum.

It is my hope that the message of this book is not lost on the adults of the children who will read this. No matter what age we are—the message of duty, love of country, personal sacrifice and courage needs to be heard over and over again throughout the ages.

— Rev. Bill McDonald Jr.

Purple Heart Veteran, Minister, Author, Poet, Artist, Documentary Film Maker, Veteran Advocate, International Motivational Speaker, Founder of "The American Author's Association" and "The Military Writer's Society of America"

A is for <u>Armed Forces</u>.

A Veteran is a person who honorably served in the Armed Forces in any of the following 13 branches: Army, Army Reserve, Army National Guard, Air Force, Air Force Reserve, Air National Guard, Navy, Navy Reserve, Marine Corps, Marine Corps Reserve, Coast Guard, Coast Guard Reserve and United States Merchant Marine.

B is for <u>Boot Camp</u>.

"Boot Camp" is a training course that transforms civilians into military personnel. Over a period of several weeks, new recruits learn about military core values, how to work together as a team and what it takes to succeed in the U.S. Armed Forces.

C is for <u>Courage</u>.

Courage is one of the many values that are needed to be successful in the Armed Forces.

D is for <u>Duty</u>.

Duty is being responsible for your own actions at all times. Our Veterans dedicated their lives to defend our country. They did their duty.

E is for the Eleventh Hour of the Eleventh Day of the Eleventh Month.

On Veteran's Day, November 11th, across our country we stand in silence for two minutes, to honor those who have served our nation.

F is for **Freedom**.

Freedom is <u>not</u> free. Sometimes you have to fight to protect it. In America, we have freedom of speech, freedom of religion, freedom to assemble and freedom of the press. Many countries don't have those precious freedoms.

G is for <u>Gold Star Parents.</u>

A parent losing a son or daughter is very difficult and stressful. If you see a gold star on a window of a home or on a license plate, it means that a son or daughter has died serving our country.

H is for Hero.

Veterans are American Heroes who should be treated with dignity, honor and respect.

I is for <u>Integration</u>.

On July 26th, 1948, President Harry Truman ended segregation in the Armed Forces by signing Executive Order 9981, stating, "there shall be equality of treatment and opportunity for all persons in the armed services without regard to race, color, religion, or national origin." Thanks to his bold step in integrating the Armed Forces, today's American military is one of the most diverse organizations in the world.

J is for <u>Joint Forces</u>.

Joint Forces are different parts of military units or branches of services, working together as one unified group to achieve specific missions.

K is for Killed in Action (KIA).

"Killed in Action" is a term that we use to describe those in the military who died at war serving our country. On Memorial Day throughout our country, ceremonies are hosted to honor those who have died.

L is for Liberty.

Liberty means to have freedom without any restrictions. Veterans have fought hard to protect our Liberty.

M is for Memorial.

In America we honor Veterans who fought to protect our freedoms and way of life. Memorials are walls, plaques, and statues that pay tribute. Memorials for Veterans can be found in our Nation's capital and in large and small communities and towns across the country.

Memorial Day is a holiday that recognizes U.S. men and women who have died serving our country. It is normally observed on the last Monday of May. Many people take time to visit cemeteries. Some people will place the American flag on each gravesite. Displaying the red poppy flower is another way to remember those who have died for our country.

N is for <u>Negotiation</u>.

History tells us there are times of peace and times of war. Most wars between countries come from controversy, conflict, and change. Being willing to negotiate and compromise are important in promoting and maintaining peace. When negotiations and diplomacy break down, it increases the likelihood of conflicts and wars.

O is for <u>Observance</u>.

On Thursday, November 11, 1954, President Dwight Eisenhower signed Veterans Day into law, making it a Federal Holiday. On Veterans Day, the President or Vice President will visit Arlington National Cemetery, and present a wreath at the Tomb of the Unknown Soldier.

Some places like post offices, banks, and libraries are closed on Veterans Day to show respect for those who served. Children, teenagers, and adults can help celebrate by sending a thank you card to a relative or neighbor who is a Veteran. Drawing pictures, writing poems, or visiting Veterans at a local VA facility or nursing home are more ways that we can celebrate Veterans Day.

P is for Prisoners of War (POW).

A prisoner of war is a service member who is captured by the enemy during or immediately after an armed conflict.

Q is for <u>Quality in Health Care</u>.

Sometimes service members get injured or ill while serving our country or after they served our country. Sometimes their wounds and scars are visible and other times they are not. In both cases, they need quality health care to live a meaningful and rewarding life.

R is for <u>Ribbons.</u>

Ribbons are worn as a badge of honor. Ribbons are similar to medals. The highest U.S. military medal is the Medal of Honor. Ribbons are displayed from the highest level to the lowest level – from left to right.

S is for <u>Salute</u>.

The Salute is a gesture by service members and Veterans of raising the hand to the cap, as a display of honor, respect, and greeting.

Service members and Veterans salute the American flag. Salutes can also be in the form of shooting a cannon or rifle. When someone in the Armed Forces dies, they will get a "21-Gun Salute" at their funeral — the ultimate honor our nation renders. Jets flying in a tight formation is another form of salute.

T is for Troops.

Troops are the men and women of the Army or Marines who normally serve on the ground. They are organized in small and large unit sizes. Air Force and Naval personnel typically don't serve on the ground, so they use different terms for their members.

U is for the <u>United States and its territories.</u>

There are veterans in every state, county, city and neighborhood. Some are young and some are old. When you meet a veteran, thank them for their service.

V is for <u>Veteran's Day</u>.

Veterans Day was first known as Armistice Day. That day celebrated the end of World War I on November 11, 1918. World War I was also called "The Great War" or "The War That Would End All Wars." Now Veterans Day (November 11th) is the day we remember and honor those who served or are currently serving in the Armed Forces.

W is for the <u>White House</u>.

The White House is the residence of the President who is also known as the "Commander-in-Chief." The President is the final decision maker of the Armed Forces.

X is for X-ray.

The military uses the phonetic alphabet to send messages by radio or telephone. For example, "Veteran" would be spelled, "Victor, Echo, Tango, Echo, Romeo, Alpha, and November."

Y is for Yellow Ribbons.

The display of a Yellow Ribbon means that loved ones are waiting for someone to come home or that someone in the Armed Forces is temporarily unable to come back home. Displaying a yellow ribbon is a practice that has been around for a long time.

Z is for <u>Zulu Time</u>.

"Zulu Time" is what the military uses to tell time. Instead of showing 12 hours on a clock or a watch, time is displayed in a 24 hour format in four digits with the letter "Z" at the end. For example, 11:06 in the morning would be written 1106Z, and pronounced "eleven zero six hours, zulu."

The End

is often the beginning

to understanding and discovery.

Acknowledgements

This is no greater time than the present to pause and give thanks to the servicemen and women, veterans, and their families who have paved our nation's future through their unselfish acts of service and sacrifice. Nearly a decade of war in Iraq and Afghanistan has proven our determination and resolve as a nation to preserve our freedoms and way of life.

Elmer Davis, the great author, news reporter, and Director of the Office of War Information during WWII once said, *"This nation will remain the home of the free only so long it is the home of the brave."* As past generations have learned and future generations will come to realize, freedom — our sacred freedom — is precious, and it is seldom free.

This book is dedicated to all who have listened and acted on the call to service in the security and defense of our nation during times of peace, and during times of war. We are eternally grateful for your courage, your contributions, and your heroism.

First and foremost, I want to thank Ms. Raqiyyah Ikram, Director, Palo Verde Kids' Club for her inspiring leadership and genuine concern as a teacher to bring outside support into her classroom to educate her students about the role of the military. It was her wisdom that set in motion the conditions that would result in me helping her students discuss and explore military culture and the significance of veterans.

I want to thank Bill Ball, Director of Volunteer Service at the Palo Alto VA Medical Center, for encouraging me to take on the challenge of going into the community to educate students about the role of our military and the contributions of our veterans.

Thanks to Maria and Jeff Edwards for their guidance and support in getting this book published.

Special thanks to the Gupta - Gnanalingam family: Gaya, Isha, Ani, and Kim for allowing me the opportunity to share in their rich reservoir of curiosity, creativity, and talents.

And finally, I want to thank my wife, Kim, and our children: Alyse, Leietta, Zora, Jonathan, Mandy, and Zach for allowing me to do what comes most naturally and fulfills my deepest passion — to be a member of the Armed Forces — the greatest military in the world, and to serve my country to the best of my abilities.

About the Editor

Gaya Gupta served as editor for this book. At age 9, her inquisitive and keen observation provided sound feedback and encouraged dialogue and discussion that greatly aided the writer and illustrator in shaping the direction of the book. Gaya's major tasks were to provide edits for age appropriateness of content and feedback for readability. Gaya lives in Palo Alto, California with her parents Kim and Ani and older sister, Isha, who served as Illustrator for this book. Gaya enjoys taking dance lessons, acting, playing violin, and hanging out with her kitties.

About the Illustrator

Isha Gupta is a gifted and talented 11 year old artist whose illustrations in this book show a deep depth of understanding and compassion for military culture and for those who have served our country. Her vivid and insightful drawings are created using several mediums that includes crayons, markers, pencils, watercolor and brush. Isha's artistic and unique ways of capturing texture and form in her drawings through vibrant colors and the use of various shades clearly reveal images, action and emotions. Her amazing ability to bring out beauty, grace, and personality in her work is seldom seen in a person her age. Isha lives in Palo Alto, California with her parents Kim and Ani and younger sister, Gaya, who served as Editor for this book. Isha enjoys learning about world affairs, taking dance lessons, and playing the flute.

About the Author

David Rabb, MA, LICSW, ACSW has been a clinical social worker for 27 years. At the age of 17, he joined the United States Marine Corps and served four years in the infantry earning the rank of Sergeant. David is currently a Colonel in the United States Army Reserves having served 23 years. He has commanded a Combat Stress Company in Iraq and is the recipient of the Bronze Star Medal. Since 1983, David has worked as a civilian for the Veterans Health Administration. He has worked with WWI, WWII, Korean Era, Viet Nam Era, Peace Time Era, Persian Gulf, Operation Iraqi Freedom, Operation Enduring Freedom, and Operation New Dawn Veterans.

David is a graduate of Illinois State University and the University of Chicago. He is married to Kim Rabb and has six children: Jonathan, Zora, Leietta, Alyse, Mandy, and Zach. He has a home in Minneapolis, Minnesota and Mountain View, California. David enjoys listening to jazz music, taking long hikes, traveling, and spending time with his family.

Portions of the proceeds from this book will go to support the Pathway Home, a non-profit residential reintegration recovery program for service members and veterans diagnosed with Post-Traumatic Stress Disorder www.thepathwayhome.org; the Coming Home Project, a non-profit organization that offers free confidential group support and stress management retreats for Iraq and Afghanistan veterans, service members, their families, and their service providers www.cominghomeproject.net; and Rebuild Hope, a non-profit organization that helps veterans and their families overcome short-term financial problems and build healthier and more stable lives www.rebuildhope.org.

Community heals. Isolation kills.

"When our perils are past, shall our gratitude sleep?"

— George Canning

For more information about veterans, learning tools, links, and other resources, visit our website at:

http://fromatozveteransbook.com

www.ingramcontent.com/pod-product-compliance
Lightning Source LLC
LaVergne TN
LVHW071028070426
835507LV00002B/61